GORILLAS

Flora Gandolfo

Grolier
an imprint of

www.scholastic.com/librarypublishing

Published 2009 by Grolier
An imprint of Scholastic Library Publishing
Old Sherman Turnpike, Danbury,
Connecticut 06816

For The Brown Reference Group plc
Project Editor: Jolyon Goddard
Picture Researcher: Clare Newman
Designers: Dave Allen, Jeni Child, Lynne Ross,
John Dinsdale, Sarah Williams
Managing Editors: Bridget Giles, Tim Harris

Volume ISBN-13: 978-0-7172-6296-0
Volume ISBN-10: 0-7172-6296-0

**Library of Congress
Cataloging-in-Publication Data**

Nature's children. Set 4.
    p. cm.
  Includes bibliographical references and
  index.
  ISBN 13: 978-0-7172-8083-4
  ISBN 10: 0-7172-8083-7 ((set 4) : alk. paper)
  1. Animals--Encyclopedias, Juvenile. 1.
  Grolier (Firm)
  QL49.N385 2009
  590.3--dc22
                        2007046315

Printed and bound in China

PICTURE CREDITS

Front Cover: **Nature PL**: Aflo.

Back Cover: **FLPA**: Paul Hobson, Fritz
Polking; **Nature PL**: Suzi Eszterhas;
Shutterstock: Emin Kuliyev.

Alamy: Blickwinkel 46; **FLPA**: Suzi Eszterhas
18, Fritz Polking 34; **Nature PL**: Ingo Arndt
26–27, Bruce Davidson 21, 38, Suzi Eszterhas
10, 13, T. J. Rich 45, Anup Shah 22, 42;
Photolibrary.com: IFA Animals 33, Andrew
Plumptre 2–3 17, 30; **Shutterstock**: Donald
Gergano 6, Eric Gevaert 9, 29, Vladimir
Korostyshevskiy 4, Hans Meerbeek 5, Steffen
Foerster Photographer 41, Ronald Van Der
Beek 14; **Still Pictures**: Fritz Polking 37.

Contents

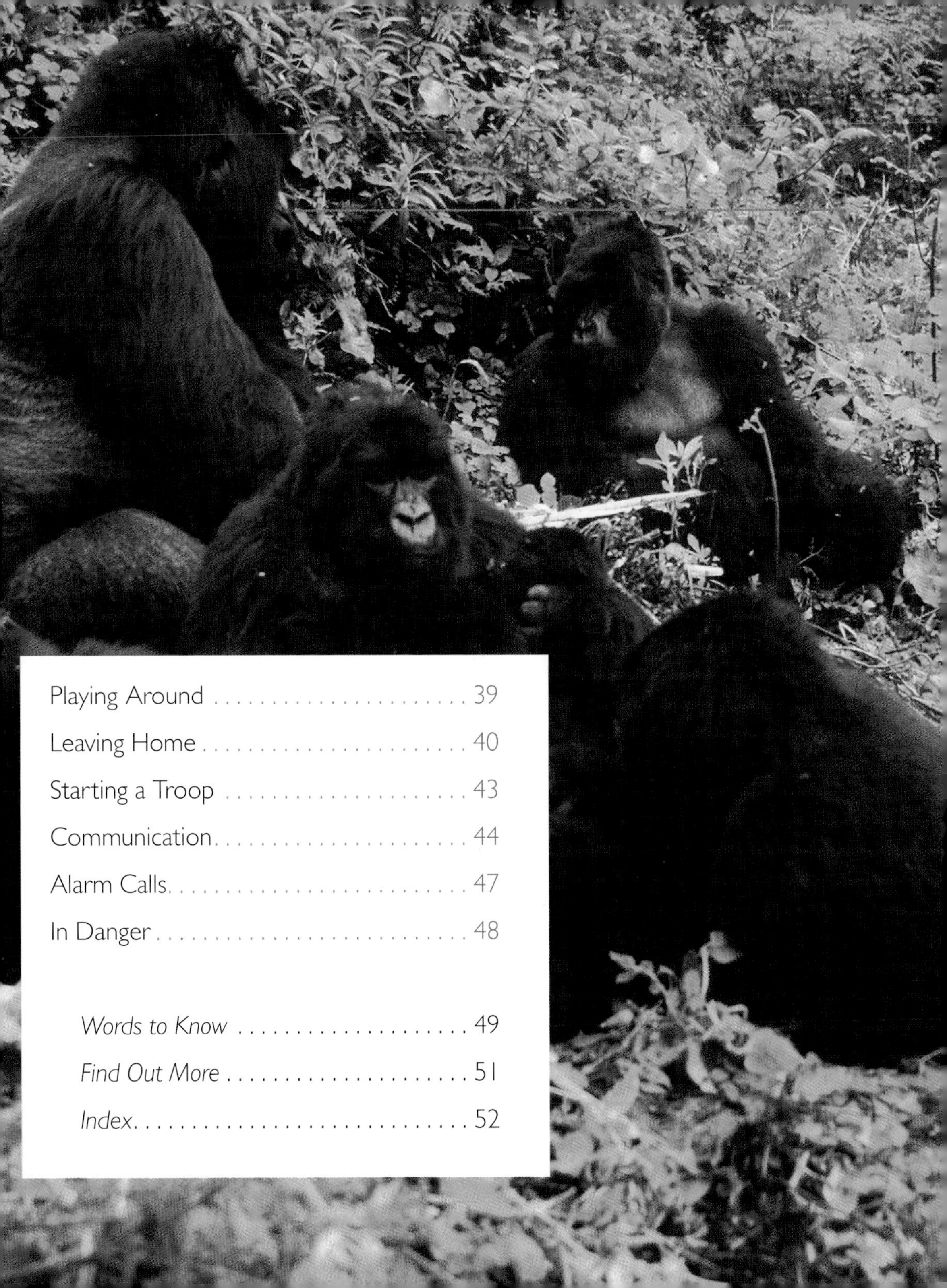

FACT FILE: Gorillas

Class	Mammals (Mammalia)
Order	Lemurs, tarsiers, monkeys, and apes (Primates)
Family	Great apes (Hominidae)
Genus	Gorillas (*Gorilla*)
Species	Western gorilla (*Gorilla gorilla*) and eastern gorilla (*G. berengei*)
World distribution	Equatorial Africa
Habitat	Forests and mountain forests
Distinctive physical characteristics	Large, muscular ape, with black skin and dark or black fur; very long arms; cone-shaped head with small ears, deep eyes, and big jaws
Habits	Live in groups; climb trees; active during the day; sleep in nests; communicate with one another by sound and body language
Diet	Fruit, vegetation, and occasionally grubs

Introduction

Gorillas are one of the closest living relatives of humans. They share many common features with humans. For example, gorillas have a large brain and are very intelligent. They communicate by using a mixture of sounds and gestures. They have hands and feet, and can stand upright. Gorillas even have individual fingerprints like humans. In fact, they are so similar to humans that when explorers first saw them, they thought they were a strange tribe of people! Their name comes from the Greek word *gorillai*, meaning "tribe of hairy women." Gorillas are large and fearsome-looking. In the past, that earned them a reputation of being violent. In reality, they are peaceful, shy, and curious gentle giants.

Gorillas are larger than all other apes—gibbons, chimpanzees, orangutans, and humans.

Male gorillas can grow to
6 feet (1.8 m) tall and weigh
500 pounds (225 kg).

The Primates

Gorillas belong to a group of mammals called the **primates**. There are more than 350 types, or species, of primates, and they include lemurs, tarsiers, monkeys, and apes. There are two families of apes: the lesser apes and the great apes. The lesser apes make up 14 species of apes. The great apes make up seven species, including chimpanzees, bonobos, orangutans, gorillas, and humans.

Zoologists—scientists who study animals—have discovered that within the great apes, humans and chimpanzees are the most closely related. Some scientists even believe that chimps should be reclassified as members of the **genus** *Homo*, alongside humans. Gorillas and orangutans are more distant relatives of humans and chimps.

Facial Features

Individual gorillas can look somewhat different, but they all share the features that give them their unmistakable gorilla look.

Their lower jaw sticks out past the upper jaw. They have large **canines**, which are the sharp, curved teeth at each side of the mouth. Gorillas also have huge muscles that control their powerful jaws. These muscles meet at the back of their head and are joined to an enormous bone that juts upward in a point. That is most clearly seen in adult males, who have a very distinctive cone-shaped head.

A gorilla's whole body is covered in thick, dark hair, except on the face, ears, hands, and feet. Just like humans, gorillas have eyes at the front of their face and small ears on the sides of their head.

Some scientists think that gorillas and humans shared a common ancestor about 8 million years ago.

Gorillas were unknown
to people living outside
Africa until about 1850.

Types of Gorillas

Wild gorillas live in the tropical forests of central Africa. There are two types, or species, of gorillas. They are the eastern gorilla and the western gorilla. As their names suggest, one species lives in eastern regions of central Africa, and the other is found in western regions.

The two species are separated by about 700,000 square miles (1,815,000 sq km) of dense forest called the Congo Basin. This area was a great lake millions of years ago. The formation of this lake is probably how the gorillas originally split into eastern and western groups. Over several millions of years the two types have evolved—changed gradually over many generations—slightly differently.

Gorillas, like humans, can look very different from one another, and yet it can be difficult for nonexperts to tell the types apart. But the two species do have differences, such as the color and length of their fur and the size of their teeth.

Eastern Gorillas

Eastern gorillas live in the uplands and mountain forests of eastern central Africa. They are found in eastern regions of the Democratic Republic of Congo, southwestern Uganda, and Rwanda. They are dark black and larger than their western relatives.

There are two kinds—or subspecies—of eastern gorillas: the more common eastern lowland gorilla and the mountain gorilla, which lives high up in mountain forests. Mountain gorillas are covered in long, silky fur to keep them warm in the cold mountain temperatures. Their fur is such a dark shade of black that it looks bluish.

There are thought to be about 700 mountain gorillas, such as this mother and baby, in the wild.

There are about
110,000 western
lowland gorillas
in the wild.

Western Gorillas

Western gorillas are much more common than eastern gorillas. They live wild in several African countries: Cameroon, the Central African Republic, Equatorial Guinea, the Democratic Republic of the Congo, Nigeria, Angola, and Gabon.

Like eastern gorillas, there are also two subspecies of western gorillas. They are called the western lowland gorilla and the Cross River gorilla. Western lowland gorillas are the most common of all gorillas and also the type usually found in zoos. There are only a few hundred Cross River gorillas in the wild.

Western gorillas have dark grayish-brown fur that is sometimes tinged with red. They also have an overhanging tip or ridge on their nose, which experts use to tell them apart from their eastern relatives.

Gorilla Troops

Gorillas live together in groups. They eat, sleep, and roam around the forest together. There are usually between 5 and 15 in a group. Some groups, however, are larger and can have as many as 40 gorillas. Adults stay in the same group for months—usually years—at a time. A group of gorillas is called a **troop**.

One adult male, usually the biggest and strongest, is the leader of the troop. He is known as the **dominant male**. He takes care of all the other male and female gorillas, protecting them from danger and leading them in search of food. A typical troop has a dominant male, a few adult females with their babies, and several young gorillas.

A troop of mountain gorillas rests and grooms one another.

Gorillas have a life span
of 35 to 40 years in the
wild, but in captivity they
can live up to 50 years.

A Day in the Life

Gorillas wake up when it gets light, which is usually about six o'clock in the morning. They immediately begin eating. The apes eat constantly until around midday. Then, when the sun is at its strongest, and it is too hot to move about, the troop gathers and rests for an hour or two.

During this time, young gorillas play around while the adults snooze. Females sometimes groom the babies and dominant male, picking out dirt and parasites, such as fleas and lice. Once they have rested, they again wander around and eat until dusk. At dusk, the animals are ready to sleep for the night.

Good Night

Gorillas find a place to settle for the night when it starts to get dark. Every time they go to sleep, they build a new nest. That keeps them pretty busy!

Nests can be in trees, on the ground, or even on steep slopes. Their nests are made from branches and leaves. They are designed to act as a cushion, protecting the gorillas from the cold, hard ground or making a snug platform up in the branches. The nests can also stop the gorillas from sliding down steep slopes or falling out of branches while they sleep.

If it is raining when the gorillas wake up in the early morning, they spend more time in their cozy nests. Gorillas hate water and avoid getting wet whenever they can.

Big males always sleep on the ground. Younger, less heavy, gorillas often make a nest in the trees.

Because the plants gorillas prefer to eat are so juicy, they rarely need to drink water.

Dinner Time

Gorillas are mainly herbivores, which means they eat plants. The treetops of the tropical forests gorillas live in are not too dense, allowing enough sunlight through for thick vegetation to grow near the ground. Gorillas live on the juicy-stemmed twigs, grasses, leaves, flowers, berries, and bark that grow on the forest floor.

Gorillas occasionally eat grubs—the young forms of insects—and adult insects, but their favorite food is fruit. That can be hard to find, but whenever the gorillas find some they devour it! Palm nuts are another favorite gorilla treat. The apes use large rocks to smash them open to get to the tasty inside.

All That Food

Because gorillas are so big, they need to spend about one-third of their day eating to get enough food and nutrients to stay healthy. Adult males can eat up to 75 pounds (34 kg) of vegetation each day. They eat their way from patch to patch, but they are clever enough to never strip an area of forest bare. That way, they can come back for another meal when the plants have grown back.

Gorillas have strong jaw muscles and large **molars**—the back teeth that grind up their food. Their large teeth and powerful jaws make it easy to process the huge amounts of food they eat.

Home Sweet Home

Gorillas do not stay in one place, but rather spend their days wandering about looking for food. They move from place to place but do not go very far, usually about half a mile (1 km) in a day. Their **range**, or the area they live in, is about 6 square miles (16 sq km). That is how much forest an average-sized troop needs to get enough food to eat.

Gorillas have an abundance of food in their range and are not territorial—they don't defend their range from other troops. A troop's range often overlaps with the range of a neighboring troop. In fact, troops don't mind mingling with other troops at all. They sometimes even spend time together if they happen to meet in the forest.

This small troop of mountain gorillas is made up of a dominant male, two females, and two youngsters.

Getting About

Gorillas have arms that are much longer than their legs. In fact, their arm span is longer than their standing height. Gorillas use both their arms and legs to move around. They walk on the knuckles of their hands, with their fingers curled underneath, and on the soles of their feet. That kind of walking is called knuckle-walking.

Gorillas are excellent tree climbers, especially when they are looking for fruit to eat among high branches. Even the enormous males manage to get up into the trees if they think they will find a tasty snack up there.

Small gorillas can also sometimes swing from branch to branch to move around. That way of moving around is common in many types of monkeys and apes. It is called **brachiation** (BRAY-key-AY-shun).

Gorillas are not the
only knuckle-walkers.
Chimps, giant anteaters,
and platypuses walk in
the same way.

29

If a dominant male silverback should die, the troop breaks up and looks for other protective males.

Silverbacks

The dominant male is the biggest and strongest male in the troop. It is easy to tell him apart from other younger males. Adult males can be almost twice the size of adult females. When male gorillas reach adulthood, they develop silvery-gray hair on their back, a bit like when humans get gray or white hair. A male gorilla's back becomes gray when he is about 12 years old. He is then known as a "**silverback**." Before he reaches this stage, a young adult male is sometimes called a "**blackback**."

The dominant silverback breeds with all the females in the troop. Because he is the only male that breeds in the group, all the other gorillas have a relationship to him—they are either a **mate** or his offspring. Young males in the troop eventually leave to find females of their own and start their own troop.

Fighting Males

Gorillas live in peace with one another most of the time, but occasionally a lone male gorilla challenges a ruling silverback. They resolve the dispute with a show of strength. A gorilla fight is very loud. The two gorillas roar, stamp their feet, and beat their chest. They tear at the foliage and charge at each other in a scary way. But instead of fighting they rush past each other, hoping that the other will be scared and run off.

They charge again and again, and that can eventually turn into a physical fight. Gorillas sometimes fight to the death, using their huge teeth to gouge deep wounds. However, more often than not the weaker of the two admits defeat and moves away from the troop. Should the challenger win, his prize is the leadership of the troop. The instinct to father new gorillas is so strong that a new leader will kill any babies that are not his own. Though brutal, that is the quickest way he can mate with the females and start a family of his own.

Gorillas have 32 teeth—the same number that humans have. However, gorillas have much bigger canine teeth than humans.

An adult female (right) rests with young gorillas. The breeding females in a troop are usually not closely related.

Family Ties

Female gorillas become adults and start to have babies of their own at about the age of ten years. Female gorillas give birth about once every four years. Not all of the babies survive. A female gorilla sometimes raises only between two and six babies during her 40-year life.

Males start to breed when they are a little older, between 15 and 20 years. A female gorilla wants the father of her babies to be the biggest and strongest silverback around. That way she knows he will be able to protect her and the babies that they have together.

Gorilla Babies

Like humans, a gorilla's pregnancy lasts for nine months. The mother usually gives birth to one baby at a time, although occasionally twins are born. Gorilla babies weigh about 4½ pounds (2 kg) when they are born. Newborns have grayish-pink skin and sparse, fluffy fur. The babies grow at a faster rate than human babies grow. They are able to crawl at three months and walk when they are less than a year old.

A baby gorilla clings to its mother's fur, so that they do not become separated. The mother carries the baby around like that for the next two to three years—until the young gorilla stops **nursing**, or breast-feeding, on its mother's nutritious milk. At three years of age, the young are no longer **infants**. Zoologists then refer to them as **juveniles**.

A young mountain
gorilla climbs onto
its father's back.

A juvenile mountain gorilla takes a rest after a bout of play.

Playing Around

Young gorillas love to play. They spend their days rolling around together, somersaulting, and chasing one another in games of follow the leader. The most important part of gorilla play is wrestling. The animals will throw themselves onto one another, hugging, clinging, and tumbling around on the ground. Sometimes, even the adults join in.

The wrestling bouts can get very noisy and are usually accompanied by loud laughterlike grunting and panting.

Playing has an important role in the life of young gorillas. It helps them to get to know the other members of the troop. It is how they learn to communicate with one another and behave within the troop. During playtime they learn behaviors that will prepare them for adult life.

Leaving Home

Young gorillas stay within the troop for many years and often stay close to their mother. They do not leave until they approach adulthood, usually between 10 and 12 years old, just as their parents did when they were the same age. The young gorillas' separation from the troop is very gradual. It starts with the young adults spending more and more time at the edge of the troop before separating completely.

Females usually wander off with another silverback and start a new troop or join his existing troop. When young males leave, they sometimes travel with other young males for several years. Eventually they find females to join them and they start their own troop.

Young gorillas have a grinning expression called a "play face" when they are having fun.

A young female mountain gorilla sits away from the rest of the troop, waiting for a lone male to come by.

Starting a Troop

A young female gorilla that is ready to leave the troop doesn't want to live on her own. She waits until a lone male is nearby and then joins him. Males lure young females away with displays of strength, by standing tall and beating their chest with their fists.

Sometimes a young female will join a group of young males, before choosing one—usually the strongest—to be the father of her babies. Once she has made her choice, she usually stays with him for life, unless he is chased away, killed by a rival or human, or dies from disease or by accident.

Communication

To communicate with one another, gorillas use a mixture of about 25 different calls, from low grunts to shrill screeches. They also use a mixture of facial expressions and gestures. Some of these are familiar to humans. They yawn when they are tired and make deep, rumbling belches when they are eating to show their appreciation!

Chest thumping is a favorite gorilla gesture. That is when they beat their chest with their fists to make a drumlike sound. It can mean many things. It can show that they are angry and ready for a fight. They also do it when they are happy, like when a person claps his or her hands together. Baby gorillas sometimes chest thump to show that they want to play.

Like humans, gorillas can tell what other gorillas are thinking and feeling from their facial expressions.

To get a message across,
this western lowland gorilla
hoots and chest thumps.

Alarm Calls

To signal danger, gorillas make sounds such as grunts, barks, screams, and roars. Male gorillas also make a hooting sound as an alarm. When the rest of the troop hears the hooting they instantly become alert.

Alarm calls often signal **predators**. Because of their size, gorillas have very few predators. Leopards and crocodiles occasionally attack gorillas, but their biggest threat is humans, who hunt them for bush meat—the meat of wild animals. In Africa, poachers, or illegal hunters, sell bush meat at a high price to ordinary people, who eat it on special occasions. In addition, many native people, who live in the same forests as the gorillas, eat bush meat just to survive.

Zoologists believe that gorillas might also communicate using smell. An adult male gorilla has pouches, or glands, in his armpits that produce a strong smell when he gets excited or scared. That could be a way of signaling to the rest of the troop that there is a threat.

In Danger

Loggers are a threat to all gorillas. Loggers destroy the gorilla's home to get wood to make a variety of products. Large areas of forest are also burned down to make charcoal, which is used as fuel in some African countries.

Wars in the areas where gorillas live have made matters worse. Gorillas sometimes get caught in the crossfire. Wars also mean that the rangers who protect the forest wildlife are not able to do their jobs because it is too dangerous. The gorillas are then even at greater risk from poachers.

The disease Ebola has recently killed many gorillas. The apes are also at risk from contact with tourists who go to see them. Gorillas are susceptible to many of the same diseases that affect humans, but gorillas have no **immunity** to them. Therefore, the gorillas are at greater risk of dying from these diseases. Fortunately, some people are working hard to protect gorillas. But there is much work to be done. Raising awareness and fund-raising are vital for the preservation of these incredible creatures.

Words to Know

Blackback   An adult male gorilla that has not yet developed the gray fur of the silverback.

Brachiation   A way of moving around in the trees by using the arms to swing from branch to branch.

Canines   The long, sharp teeth at the sides of the mouth in mammals.

Dominant male   The male gorilla that is the leader of a troop.

Genus   A group of closely related species.

Immunity   Being able to resist a particular disease.

Infants   Baby gorillas up to three years old.

Juveniles	Young gorillas between three and six years old.
Mate	Either of a breeding pair; to come together to produce young.
Molars	Large back teeth that grind up food.
Nursing	Drinking milk from the mother's body.
Predators	Animals that hunt other animals.
Primates	A group of mammals that includes lemurs, tarsiers, monkeys, and apes. Primates have a large brain, eyes that face forward, hands, and feet.
Range	The area that a troop of gorillas lives in.
Silverback	An adult male gorilla.
Troop	A group or family of gorillas.

Find Out More

Books

Dennard, D. *Gorillas*. Our Wild World. Minnetonka, Minnesota: NorthWord Books For Young Readers, 2003.

Turner, P. S. *Gorilla Doctors: Saving Endangered Great Apes*. Boston, Massachusetts: Houghton Mifflin, 2005.

Web sites

All About Gorillas
www.enchantedlearning.com/subjects/apes/gorilla/
Tons of facts about gorillas.

Creature Feature: Mountain Gorilla
kids.nationalgeographic.com/Animals/CreatureFeature/ Mountain-gorilla
Information and pictures of mountain gorillas.

Index